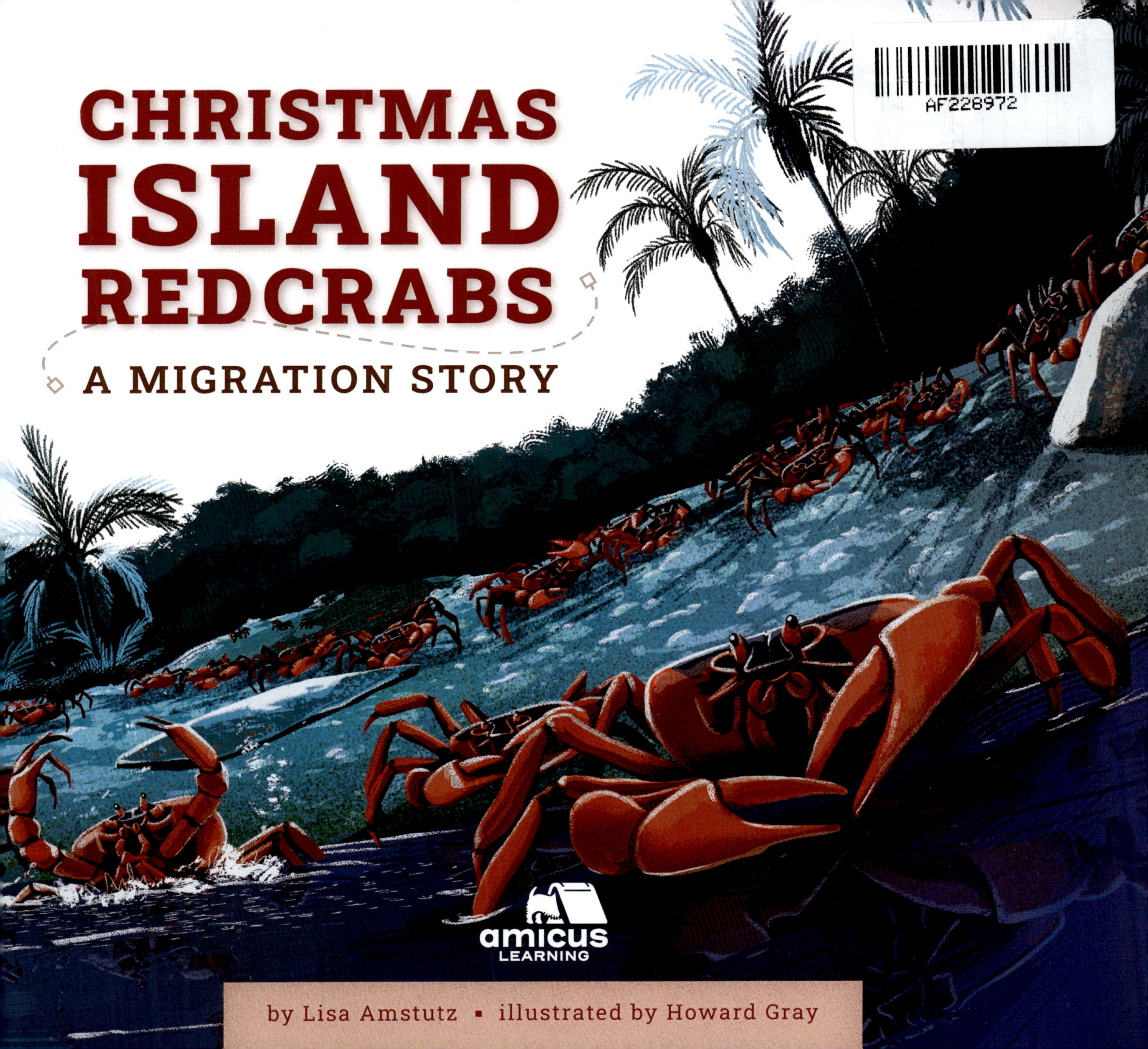

CHRISTMAS ISLAND REDCRABS

A MIGRATION STORY

amicus
LEARNING

by Lisa Amstutz • illustrated by Howard Gray

AMICUS ILLUSTRATED is published by
Amicus Learning, an imprint of Amcius
P.O. Box 227, Mankato, MN 56002
www.amicuspublishing.us

Editor: Alissa Thielges
Series Designer: Kim Pfeffer
Book Designer: Emily Dietz

Library of Congress Cataloging-in-Publication Data
Names: Amstutz, Lisa J., author.
Title: Christmas Island red crabs : a migration story / by Lisa Amstutz.
Description: Mankato, MN : Amicus Illustrated, [2026] | Series: Incredible migrations | Includes bibliographical references. | Audience: Ages 5–10 | Audience: Grades 2–3 | Summary: "Follow the incredible journey of a Christmas Island red crab's migration in this narrative nonfiction picture book that will delight animal and nature lovers and support life science education. Includes a migration map, tips to protect animals and habitats, a glossary, and further resources"— Provided by publisher.
Identifiers: LCCN 2024043358 (print) | LCCN 2024043359 (ebook) | ISBN 9781645492818 (library binding) | ISBN 9781681528052 (paperback) | ISBN 9781645493693 (ebook)
Subjects: LCSH: Gecarcoidea natalis—Juvenile literature. | Gecarcoidea natalis—Migration—Juvenile literature.
Classification: LCC QL444.M33 A537 2026 (print) | LCC QL444.M33 (ebook) | DDC 595.3/86—dc23/eng/20250115
LC record available at https://lccn.loc.gov/2024043358
LC ebook record available at https://lccn.loc.gov/2024043359

About the Author
Lisa Amstutz is the author of more than 150 children's books. A former outdoor educator, she holds degrees in biology and environmental science. Lisa enjoys learning fun facts about science and sharing them with kids. She lives on a small farm with her family.

About the Illustrator
Howard Gray has illustrated a selection of fiction and non-fiction children's books. He has always considered himself an artist, but with a PhD in dolphin genetics, he has a background in zoology. He is now pursuing his dream career in children's illustration from the picturesque city of Durham, UK. Find out more at www.howardgrayillustrations.com.

Skitter, skitter! A red crab snatches a piece of fruit. She returns to her burrow to eat it. The crab lives on Christmas Island in Australia. Millions of red crabs make homes on the forest floor. They eat leaves, fruit, and flowers.

The crab pulls a leaf over her door. She hides from the sun. It is the dry season now. The hot air dries out her gills and makes it hard to breathe. But soon, the fall rains will come. Then she will begin her journey to the sea. She must find a mate and lay her eggs.

Splish, splash! The rains begin. The air is moist. It is time to migrate. The crab joins millions of others. The journey can take 18 days. The crabs climb hills and tumble down cliffs. Nothing can stop their march to the sea.

Yellow crazy ants swarm the crab. The ants kill thousands of crabs each year. This crab shakes them off, but others are not so lucky.

Another danger ahead! The crab must cross the road. But people
help. They stop traffic. They make bridges and tunnels for the crabs.

The crab speeds up. She must hurry! The shore is still miles (kilometers) away. She needs to arrive a few days before the full moon. That way she can lay her eggs at the right time.

At last, the crab reaches the shore. Her gills are dry. She takes a dip in the ocean to wet them.

The male crabs are waiting. They left a few days early to dig burrows. The female chooses a mate. She settles into the hole that he made.

The male takes a dip in the sea and heads home. His job is done. The female stays behind. She lays up to 100,000 eggs. She holds them in her brood pouch.

After 12 days, the crab heads for the water.
She finds a shady spot to wait.

The moon is in its last quarter. Just before dawn, the high tide retreats. The crab dashes into the sea to spawn. She bobs up and down in the water. Her eggs float away.

Crab eggs cloud the water. Soon larvae will hatch out. They will float in the ocean for a month. They will molt and change form. Many will be eaten by fish or manta rays. But some tiny red crabs will return to Christmas Island.

The female crab's job is done. Now it is time to go home.
She makes her way up the sandy shore.

The crab walks for days.
At last, she reaches the forest.
She finds a burrow and moves
in. Now she will feed and rest
and hide from the sun . . .

until the rains come again.

CHRISTMAS ISLAND RED CRAB MIGRATION

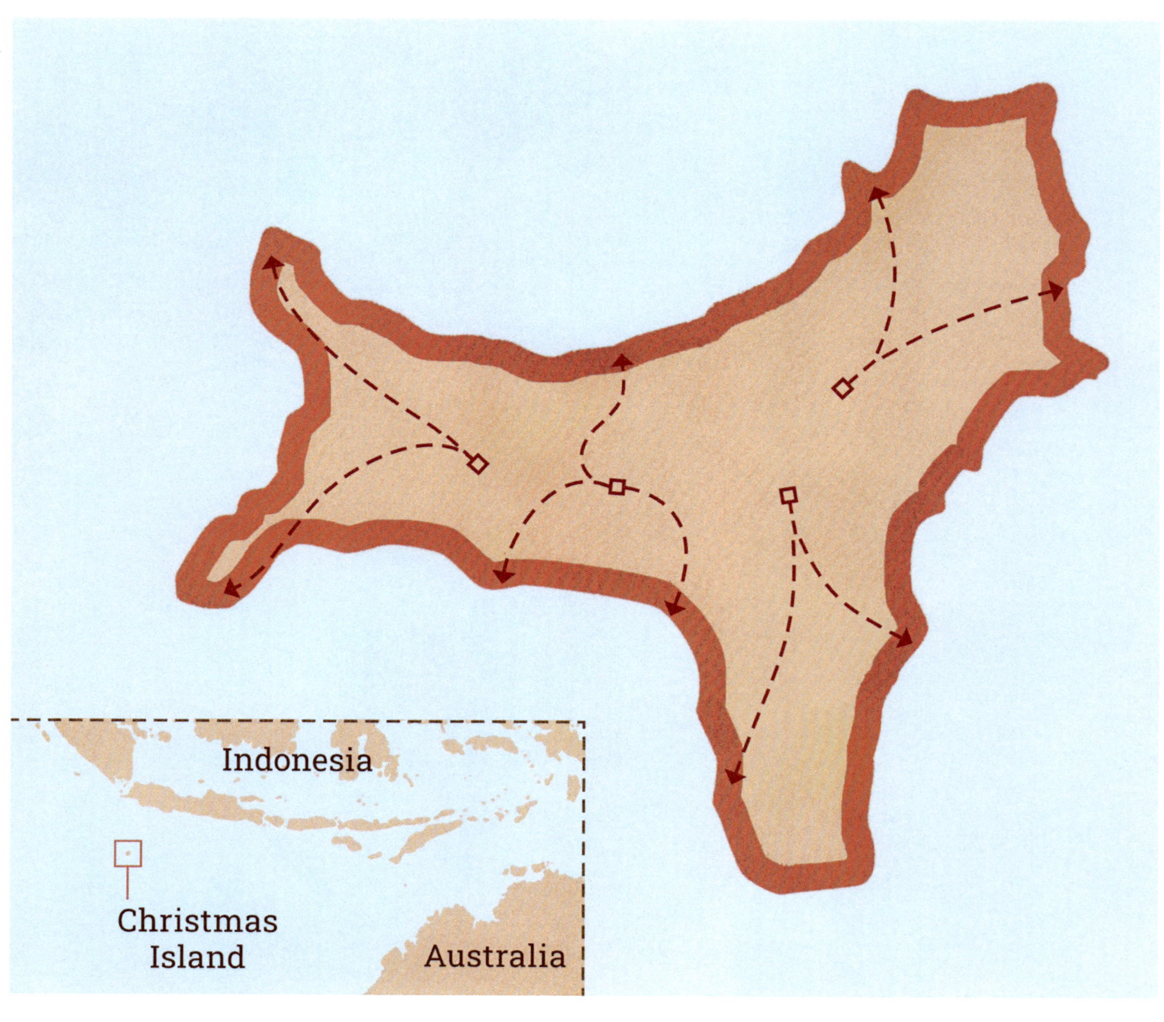

Humans are trying to help red crabs migrate safely. You can help by keeping the ocean clean. These tips can save animals in your area, too.

- Throw away trash and pick up litter to keep nature clean.

- Use reusable water bottles and bags rather than plastic ones. Plastic can end up in the ocean and harm wildlife.

- Use less water so your wastewater doesn't flow into the ocean.

Glossary

brood pouch A pouch that protects the crab's eggs before hatching.

burrow A hole or tunnel dug by an animal.

gill An organ that some animals use to breathe underwater.

larva, *plural* **larvae** A stage of development that comes after the egg, but before transforming into an adult.

mate The female or male partner of a pair of animals.

migrate To move to another area at a certain time of year.

molt In crabs, to shed the hard, outer shell after the crab gets too big for it; a new shell then grows.

spawn To lay eggs.

Read More

Banks, Rosie. *Why Do Animals Migrate?* New York: Gareth Stevens Publishing, 2024.

Brundle, Harriet. *Animal Migrations.* Minneapolis: Jump, Inc, 2024.

Donnelly, Rebecca. *On the Move with Christmas Island Red Crabs.* Minneapolis: Jump!, Inc., 2023.

Websites

Behind the Scenes: Red Crab Migration
https://thekidshouldseethis.com/post/63024390300

Christmas Island Red Crab
https://kids.nationalgeographic.com/animals/invertebrates/christmas-island-red-crab/

Red Crab Migration
https://parksaustralia.gov.au/christmas/discover/highlights/red-crab-migration/